Praise for *On The River*

On The River is a great read for both the budding and reluctant scientist in your life. Readers can easily follow along with Max and Ellie's river adventure and have one of their own! And if all the great biology facts throughout the story aren't enough, the Trail Guide at the end will be sure to spark even more curiosity. The heart of the book shines through clear: anyone can do science!

—Renee Becher, owner of The Crowded Bookshelf

I liked how this book was scientific. I want to take this book to my river and join Ellie and Max!

—Evangeline Lillie, age 5

This book inspired me to challenge myself when there is something I don't think I can do. *On the River* is the perfect book for kids like me who are interested in science and learning about the world around them.

—Morgan Morehouse, age 9

This book takes you on a scientific adventure that you can replicate along a river in your town! I love the details that can guide young kids to determine if their river is healthy.

—Sarah Lillie, elementary school teacher

On the River celebrates the joy of childhood adventures and outdoor discovery. This beautifully crafted story presents science exploration in a way that's exciting and allows young readers to recognize that they can also be scientists! Throughout the story, readers will be reminded to challenge beliefs about their own abilities to engage with science. The illustrations and trail guide will delight readers both young and old alike!

—Sara Piotter, Senior Environmental Educator, Environmental Learning Center

Most of us think, at one time or another, that we're no good at something, especially if that something is unfamiliar. *The Biology Adventurers: On the River*, sparks a little bit of curiosity and offers friendly encouragement that **no good** can be replaced with **pretty fun!** With accurate names and descriptions of the biological world, but a playful sense of adventure, I was reminded that biology begins right where my own toes end.

—Jessica Miller, Education and Outreach Coordinator, Rocky Mountain Raptor Program

THE BIOLOGY ADVENTURERS

ON THE RIVER

Written by Melanie Peffer

Illustrated by Valentina Gallup Salerni

Publisher's Cataloging-in-Publication data

Names: Peffer, Melanie E., author. | Salerni, Valentina Gallup, illustrator.
Title: The biology adventurers : on the river / written by Melanie E. Peffer, PhD; illustrated by Valentina Gallup Salerni.
Series: The Biology Adventurers
Description: Greeley, CO: MKPEF4, 2021. | Summary: Ellie wants to earn her Biology Adventurer River Badge with her brother Max. But Max doesn't think he's any good at science. Can Max overcome his nerves and earn his adventurer badge, too?
Identifiers: LCCN: 2021922436 | ISBN: 978-1-7346531-3-7 (hardcover) | 978-1-7346531-4-4 (paperback) | 978-1-7346531-5-1 (ebook)
Subjects: LCSH Rivers--Juvenile fiction. | Outdoor life--Juvenile fiction. | Brothers and sisters--Juvenile fiction. | Ecology--Juvenile fiction. | Nature--Juvenile fiction. | Nature stories. | BISAC JUVENILE FICTION / Nature & the Natural World / General | JUVENILE FICTION / Science & Nature / Environment
Classification: LCC PZ7.1.P4442 Bio 2021 | DDC [E]--dc23

M.P.: To everyone who has ever felt like Max.

To Felix – who inspired many of the scenes in this story

V.G.S.: For Italo and Leo

ACKNOWLEDGMENTS

Thank you to the High Plains Library District Foundation for financially supporting this project through the Writer in Residence Program.

Thank you to Ambrette Gilkey and Larry Rogstad from the Poudre Learning Center for help getting the science right.

Big thank you also to members of the Boulder Writers Association and Northern Colorado Writers for support and critiques of early versions of the manuscript.

Max was working on building a model airplane when his door burst open, and his big sister Ellie walked in.

"Hey, Max! Want to go to the river? I need to figure out if the river is healthy to earn my Biology Adventurer River badge," she said. "This is my last badge before I can advance to a Biology Adventurer Leader."

Max groaned. Ellie was super smart. She loved biology. She was always rattling off random facts.

"Another badge, Ellie? Don't you have enough? The kids at school already make fun of you for being a science geek."

"Science geek is a compliment, Max," she said. "If you earn your river badge today, you can advance to level one. Then you'll be ready for your first real science class next school year."

"I'm not sure, Ellie. I'm not good at science. It's too hard."

"Why do you think science is hard?"

"I heard it from the kids at school," he said. *Besides*, he thought to himself, *pilots don't need to know about science*.

"How do you know if you don't try? It can't be more difficult than learning how a build a plane. Besides, the model airplane club meets at the river on Saturdays."

Max loved seeing all the different kinds of planes at the airplane club. Maybe that would make doing science with Ellie tolerable. "Okay." He sighed. "I'll go with you."

"To earn the Biology Adventurer River badge, we'll need to look for signs of health in and around the water, plus how well it does its job."

Max was already confused, and they hadn't even made it to the river. "How does a river do its job? What does it even mean, for a river to do a job? And how is a river healthy? Does it go to the river doctor for a checkup?"

Ellie giggled. "Rivers don't go to the doctor. But they can be healthy or sick, like our bodies. A river's job is to filter sand and nutrients and provide a home for plants and animals."

"Right ... so what is the first sign we'll look for, then?"

Ellie smiled broadly and held up her Biology Adventurer book. It was her most prized possession. The book had information about different ecosystems and what kinds of plants, animals, and bugs were found in each place.

“We look in the water for bugs!”

Max groaned. He hated bugs. He wished he was back at his bug-free desk working on his model airplane. He hoped he hadn’t made a mistake by saying yes.

There was a clear blue sky overhead, and the sun twinkled merrily on the water. Max was searching the sky for model airplanes when Ellie plopped a bucket of water down in front him, splashing him.

As Max shook off the water, he peered into her bucket.

“Why is the water wiggling like that? Is there a creature?” Max was worried a bug would jump out at him.

Then Max saw it.

“What is that?”

“The book says to look at its tail,” said Ellie. “I see three tails, so this is either a mayfly or a damselfly nymph.”

“But it doesn’t look like a fly! It looks like a leech.” Leeches can’t fly, but they do suck blood. Max was starting to feel sick to his stomach.

“Nymphs are baby insects,” said Ellie. “They’ll fly when they are grown up. Now, a damselfly nymph has a paddle tail, and a mayfly nymph a string tail. Which one do you think it is?”

"I see three strings ... so it's a mayfly nymph!" said Max excitedly. Identifying the nymph made Max forgot his queasiness. "But Ellie, what does finding a mayfly nymph have to do with the river's health?"

"Mayfly nymphs can't live in water with pollution. It says here that you only find mayfly nymphs in water that is clean. This is a good sign the river is healthy."

"Great, the river is healthy. Can we go home now?" Max searched the sky for any sign of the model airplanes. Something bright yellow shot overhead. But it wasn't a plane. And it made a *sweet sweet sweet* noise.

"Look!" Ellie said. "It's a yellow warbler! Looking outside the water is the next thing on the checklist. Let's see, it's says here that yellow warblers are a sign of a healthy river. They will only nest in willows alongside rivers that have lots of bugs."

Hmmm, thought Max. This didn't seem like science at all. He liked putting the pieces of evidence together, just like one of his model airplanes.

"If we found a mayfly nymph, it means the river isn't polluted, so there should be lots of bugs around ... and lots of food for birds," Max reasoned. "I think this is another sign that the river is healthy."

"I think you're getting the hang of this, Max!"

"Okay," she said. "We've looked in and around the water. The last thing on the checklist is to evaluate if the river is a continuous road. It says here that rivers are like highways for animals to move from one place to another."

"It's all one river. Why wouldn't it be one long highway for animals?" he asked.

Ellie had that gleam in her eye, the one she got when she was excited. "Do all animals walk?"

Something clicked in Max's brain. Airplanes didn't need to drive on roads. They could fly—the same as birds.

"The yellow warbler. It lives near the water, in willows, not in the river." Max glanced at the houses along the bank of the river. "I don't see any willows," he said. "Even if there are lots of insects here, the yellow warbler wouldn't have anywhere to build its nest."

Ellie furiously scribbled notes in her book. "The houses are a barrier to the river highway. This is a sign the river can't do its job as well."

“Ellie, look over here! Is this tree hurting the river highway? Wouldn’t fish have trouble swimming around it?”

“The book says that downed trees are good for the river. They provide structure for small animals,” said Ellie.

"Like what?" Max was surprised when the question rolled off his tongue. Where were all these science questions coming from? Pretty soon he was going to start sounding like Ellie.

"Let's go look!"

They splashed into the water near the fallen tree. Max jumped back with a loud yell. "It's a lobster!"

"A what?" said Ellie. "Lobsters live in the ocean!"

"Look for yourself!" Max answered.

Ellie leaned down and scooped up the lobster, excitedly yelling, "It's a crayfish! Crayfish keep the river clean. They're also food for birds, fish, otters, and even humans."

"People eat these? No, thank you!" Max laughed. "But crayfish seem important. So finding a crayfish must be another sign that the river is healthy."

Under the shadow of the tree, Ellie spied a group of fish. "Why are some of the fish smaller? Is the big one the mother, and those are all her babies?"

Max grabbed the Biology Adventurer book out of Ellie's hands and flipped to the fish page.

"The big one looks like a carp. Carp are an invasive species. The little ones are shiners and minnows."

Ellie leaned over Max's shoulder to look at the book. "If it's invasive, what is it doing here?"

"Maybe it's visiting Earth from the Carp Mother Planet." Max laughed.

Ellie rolled her eyes and pointed to a line in the book. "An invasive species is one that isn't supposed to be here but was introduced either accidentally or on purpose. Carp were introduced by immigrants for food. Different continent, not planet, Max!"

"It says here that when carp eat, they dig around on the bottom of the river. This causes the water to get cloudy. Cloudy water is unhealthy. Carp also eat the other fish's babies."

INVASIVE SPECIES? -
- YES OR NO?
QUALITY

Then Ellie frowned. "But Max, I'm confused. I'm not sure if the river is healthy or not. We found some signs of health, like the crayfish and the warbler. But what about the houses along the river and the carp? It can't be healthy if there are carp here. I don't know what the answer is. There has to be an answer. Isn't that what science is all about? Finding answers."

As Ellie talked, she sounded more and more frustrated. Max knew that Ellie prided herself on all the science stuff she knew. Then Max thought about airplanes, how different kinds of airplanes are suited for different purposes. There wasn't necessarily a right or wrong airplane to use, but one that fit a situation better. Could science be the same way?

"Maybe science doesn't always give us an exact answer," he said. "It's like why there are so many different types of planes. We have to look at what's going on, like the evidence, and decide what fits best." Max pointed at the book. "It even says here we have to look at all of our evidence and make a decision. There isn't necessarily one answer. If a river's health means how well it works, then I think our river is functioning well. If we keep an eye on the signs listed on the checklist, we can keep the river healthy."

Ellie breathed a huge sigh of relief. "Wow, Max, you really helped me earn my Biology Adventurer Badge today! Do you still feel like you aren't good at science?"

"Being good at science isn't what I thought it was," he admitted. "It's not just about knowing a bunch of science stuff like you, Ellie. It's about doing and thinking. I had fun collecting evidence today and connecting the pieces together." It was a good day, even if he didn't see any model airplanes.

"What will our next adventure be, Ellie? I'm ready to go!"

TRAIL GUIDE

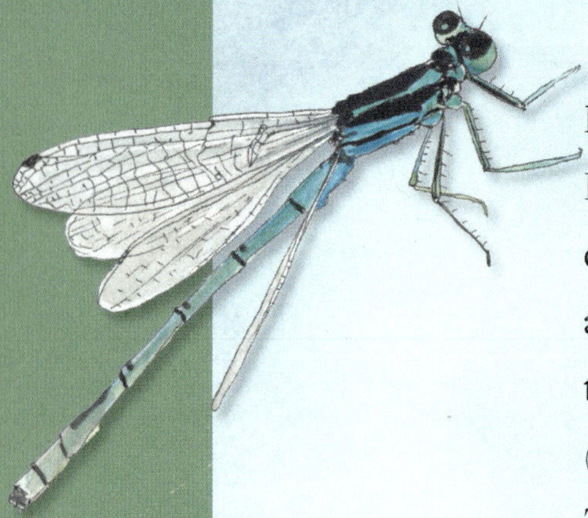

DAMSELFLY: Damselflies look like dragonflies but are different insects. Dragonflies hold their wings away from their bodies while resting while damselflies hold their wings close to their bodies. Damselflies are an ancient species, first observed over 250 million years ago. They survived two mass extinction events, including the Cretaceous-Tertiary event (K-T event) that wiped out the dinosaurs. Damselflies are predators. They eat other insects. Baby damselflies or nymphs only live in unpolluted, fresh water. Seeing damselflies near a river indicates that the nearby water is clean.

MAYFLY: Mayflies are also called "up-winged flies" because they hold their wings upright like a butterfly when at rest. Mayflies have been around as long as damselflies. Have you ever wondered what the first insects looked like? Mayflies give us some clues, with their long tails and upright wings. Like damselflies, mayflies are only found near clean, unpolluted water. Their presence indicates good water. Mayflies go through three stages: nymph (aquatic), subimago (terrestrial, with some flying), and imago (airborne).

LEECH: Leeches are a kind of worm that lives in fresh water and feeds on blood. Be careful wading in fresh water, or you may pick up some of these parasitic worms! Leeches aren't all bad, though. Doctors have used leeches for medical treatment for a long time. Leeches make chemicals to keep blood flowing. This can help prevent dangerous blood clots from forming. Leeches live in polluted or unpolluted water, so they aren't a good measure of clean water.

MOSQUITOES: These flying, bloodsucking insects are among most humans' least favorite insect! In fact, these tiny insects cause more human deaths than any other organism on the planet. Why? Because when mosquitos bite, they can transmit deadly diseases such as West Nile, Zika, or yellow fever. However, although we might find mosquitoes annoying, they are an important part of a healthy river. Mosquitoes, in their aquatic larval form or adult flying form, are an important food source for other insects, mammals, birds, reptiles, amphibians, and fish.

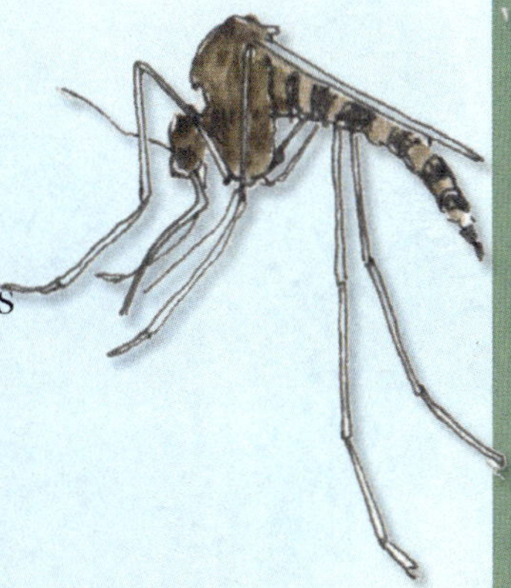

YELLOW WARBLER: Yellow warblers are bright yellow songbirds that live near trees or shrubs along a body of water. Their bill is long and thin, which helps them catch their prey—insects! The males sing a characteristic *sweet sweet sweet* song. Birds sing to communicate. One reason warblers since is to attract a mate. Scientists study how birds communicate through song. This helps them understand human speech and the development of language disorders better.

OSPREY: Ospreys are large predatory birds, or raptors. Sometimes people call ospreys "fish hawks." Ospreys carry fish in a special position so they can fly more aerodynamically. If you watch carefully, you might see an osprey toss the fish to get it into position! Ospreys signal a healthy body of water because they only live where there are lots of fish. Ospreys also represent what can happen if humans are good stewards of natural places. For example, people build osprey nesting platforms for these magnificent birds and live in harmony with their avian neighbors.

GREAT BLUE HERON: Like the osprey, the presence of these predatory birds tells us that a river has lots of fish and is likely healthy. Although herons are more adaptable than other birds, they typically wade in clean, healthy aquatic environments. Herons can be found in fresh and saltwater environments. Look for a characteristic S shape in their neck as they fly.

BELTED KINGFISHER: Belted kingfishers are shy and can be hard to spot. However, their presence is an excellent sign of river health, as they only live near water with not only many fish, but many kinds of fish. Look in the branches of downed trees near or over the water for this small bird with its characteristic fluffy head and long bill. Watch long enough, and you may see the kingfisher dive into the water for a treat!

SANDBAR WILLOW: Sandbar willows are a type of shrub that can grow up to twenty-three feet tall. The leaves are long, skinny, and green. They are deciduous, meaning they lose their leaves in the fall. Sandbar willows flower in the spring and are important to many pollinators, including many butterfly species. Native Americans built home with willow branches. Willow bark is a folk remedy for cough and sore throats. Willow bark contains a chemical called salicin. Salicin changes to salicylic acid (the active component of aspirin) when eaten.

NARROWLEAF WILLOW: Narrowleaf willows are found throughout the United States next to rivers such as the Yukon in Alaska and the Mississippi in Louisiana. It is also a type of shrub and can grow up to fifteen feet tall. Like sandbar willows, narrowleafs also bloom in the spring and are important to pollinator species like butterflies, moths, and bees. Moose and elk also enjoy munching on willows.

COTTONWOOD: Cottonwood trees grow alongside rivers and can reach one hundred feet tall. They also stabilize stream and riverbanks and aid in filtering rivers, promoting good water quality. Cottonwoods get their name from the cotton-like seeds produced by female trees. Cottonwoods have brittle branches which can lead to property damage when planted near homes, but they can also supply a source of structure within rivers or streams and a welcome hiding place for fish and other aquatic creatures.

CRAYFISH: Crayfish (also called crawdads, crawfish, mudbugs, and mountain lobsters) are a freshwater crustacean related to ocean-dwelling lobsters. Crayfish are a keystone species, meaning that many other organisms depend on them. Crayfish help keep water clean and are an important source of food to a variety of other animals, including humans. Crayfish are extremely sensitive to pollution. In fact, sometimes they are used to let humans know if there are changes in water quality.

PLAINS MINNOW: Like other fish native to rivers in dry environments, they are very tiny in size, only about five inches (thirteen centimeters). Since rivers in dry environments can become very shallow during the summer months, the native fish evolved to be small to survive in shallow water. plains minnows are threatened by non-native fish like walleye and carp.

CARP: Carp are a type of fish native to Europe and Asia. Carp were introduced in the United States by immigrants in the 1800s. Carp breed prolifically and can survive in polluted water, which facilitated their rapid expansion in the United States. In addition to outcompeting native fish, they dig around river bottoms to eat insect larvae and uproot plants, which disturbs the ecosystem and decreases water quality. Fishing carp for human consumption may be one of our best bets for preserving waterways in the United States.

HOUSING DEVELOPMENT: Like many other animals, humans enjoy living on or near bodies of water. Expansion of human housing along rivers removes critical habitat for other animals. However, humans can be good stewards by making efforts to preserve quality habitat alongside the river or include structures, such as nest platforms for osprey. We have many chances to act as good stewards of nature and figure out how humans fit into ecosystems.

IRRIGATION DITCH: Irrigation ditches are one way humans have changed rivers to suit their needs. For example, in places like Colorado that are drier, ditches were built to harness water from the river to use for growing food. Thanks to the ditches and deepening the river, fish such as carp or walleye successfully live in waters that were traditionally shallow in the summertime. Native plant species are often removed during construction of ditches.

Author's Note

Since I first started teaching biology in 2007, I've noticed repeatedly how many people—including my students, friends, and family members—feel like they can't do science. That it's too hard. That they aren't smart enough to understand it. Even the early readers of this story talked about how much they identified with Max's character.

I founded the Biology Everywhere project around the idea that anyone can engage with biology. That biology is all around us and isn't about knowing a bunch of facts. My philosophy is that when we see biology everywhere, we engage with it—and once we begin to engage, we realize that we are smart enough to understand and that it is also more interesting than we ever realized. Starting with the small things that we see every day is a steppingstone that eventually leads to the confidence to engage with bigger science issues in society.

I specifically crafted this as a children's story, because research suggests that late childhood is the period when children begin to feel less like a "science person," particularly among girls. I also find pictures books to have universal appeal regardless of age. My hope is that all readers, and especially children, can identify with either Ellie or Max, follow them on their journey, begin to see biology everywhere, and feel like a "science person." That, like Max, they realize that science is more than memorizing a bunch of abstract facts. Or like Ellie, realize that they can be a role model.

The scenery and life in the book depict the Cache la Poudre River, which flows through my hometown in Greeley, CO. However, you will find many of the organisms depicted in the book near or in any river. I encourage you to visit your local river with a child see what you find in and along the river. Do you think your river is healthy? Why or why not? Come tell me about it on social media:

Twitter: @Melanie_Peffer

Instagram: @biologyeverywherebook

Facebook: www.facebook.com/melaniepefferphd

For more information on Biology Everywhere, including ancillary activities and updates on new books visit www.biologyeverywhere.com

About the Author

Dr. Melanie Peffer has a B.S. and Ph.D. in molecular biology from the University of Pittsburgh. She is on faculty at the University of Colorado Boulder, where she teaches introductory molecular biology courses and studies how people learn biology, with an emphasis on teaching about the nature of science knowledge and how we can use educational technologies to assess students' understanding of science.

Dr. Peffer is an in-demand speaker and educator, regularly speaking at both national and international venues on topics including her research, teaching science, and writing for the general public. She has collaborated with TED-ED to produce a science lesson, and gave a TEDx talk in 2021.

Dr. Peffer is the author of the best-selling book, *Biology Everywhere: How The Science Of Life Matters To Everyday Life*. *Biology Everywhere* is a journey through the science of life as told through our daily experiences. She was also the 2021 High Plains Library District Foundation's writer in residence. When not writing, Dr. Peffer enjoys playing her flute and piccolo and enjoying all that Colorado has to offer in the great outdoors with her husband and son.

About the Illustrator

Valentina Gallup Salerni is an interdisciplinary artist with a background in painting and drawing and enthusiasm for creating educational and visual work featuring native species and their surroundings. She is the illustrator of *Horseshoe Crab's Crown*, and contributes artwork regularly to local cultural and environmental organizations.

Believing art and science are undoubtedly intertwined her work hopes to nurture imagination and curiosity in children of all ages.

Valentina is a New York City native, currently living in Rockaway Beach, New York, where she is continuously inspired watching her two young children engage with the natural world.

Made in the USA
Monee, IL
22 November 2021

82426637R00026